Connect with Nature books

THE SEASONS AND ME

a story about the yearly body clock

by Suzanne Mathis McQueen

Illustrations by Pumudi Gardiyawasam

Biology ~ Nature ~ Health
Paperback ISBN: 978-0-9858955-7-0

Special thanks to:

Ask Janis Editorial Services
Pumudi Gardiyawasam, illustrator extraordinaire!
Patrick Wagesreiter, permission for Patrick Hand font
Ms. Galaxy, Jessica Vineyard
Rocco and Jazzi (the dogs)
George, Elmer, Mabel, and Chico (the cats)
Blaine and Erin Moore of Apex Authors

Superstar Beta Book Readers:
Myan McQueen, Louise Abel, Cynthia Salbato, Laura Knapp,
Jane Almquist, Linda Parker, Sofia Henry, Sara Parker,
Judith Clark, June Kenney-Clark, Eliza Kenney-Clark,
Jeanie Vollrath, Tracie Smith, Mark Siino, and Aurora Perez

Four Seasons Song by Jack Hartmann
https://jackhartmann.com

Let's sing the Four Seasons song
Winter, Spring, Summer, Fall
Each year has 4 Seasons in all
Winter, Spring, Summer, Fall

by Jack Hartmann

Four Seasons Song

For my kids, Ian, Preston, and Myan.

Here's to our great Halloween porch and all our fun times in Winter, Spring, Summer, and Fall.

The Sun, the Moon, and the Seasons
are magical timekeepers!

Our bodies follow the Sun's daily clock,
the Moon's monthly clock, and the Four Seasons' yearly clock.

This is the story of the Four Seasons' clock and me!

Mother Nature has four children.

They are known as the Four Seasons.
Their names are Winter, Spring, Summer, and Fall.

One at a time, each Season steps in to say "Hello!"

They take their turns at guiding us
through the year.

The Seasons bring us different gifts
in the form of light, food, plants, weather,
projects, and fun adventures!

Throughout the year, the Seasons teach us how to live with Nature.

They let us know when to wear a coat and when to wear shorts and a T-shirt.

The Seasons tell our bodies what
time of the year
we need to sleep longer...

...and when we can stay out
longer to play.

The Seasons have their own personalities!

Depending on where you live, they can be brave and fierce, or quiet and shy, or a little bit of both.

I'll show you what the Four Seasons look like where I live!

Winter is cold and wild!
It likes to howl with the wind.

Winter's job is to create a lot of snow in the mountains so that when it melts in the Spring and Summer, there will be plenty of water in the lakes and rivers.

Winter's birthday is called Winter Solstice.

It's the longest night and shortest day of the year.

Winter Solstice activates the RETURN OF LIGHT!
Each day gains a bit more Sun time as we move toward Spring.

As each day builds, my energy grows!

Winter is magical.

It's fun to build a snowman.

Hot chocolate with
marshmallows is the BEST.
Yum!

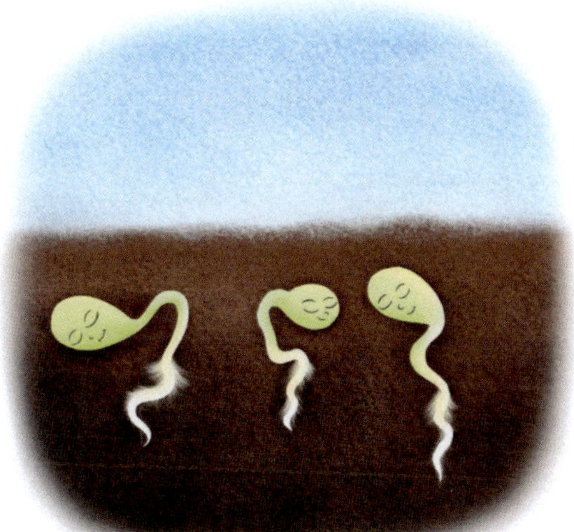

Under the ground, seeds hide
from the cold and then begin to
wiggle and sprout.

My town creates a Winter Solstice Spiral out of river rocks or greenery. It's very pretty!

Each person is given an unlit candle.

Then, as we walk silently toward the center, we think about the good things that happened during the year and give thanks.

When we get to the center, the candle is lit to celebrate the return of longer days!

When following the path out, we think about the good things we want to create in the upcoming year.

After three months, Spring takes over from Winter.

Even if the weather still feels like Winter,
Spring will wave its magic wand to BOOST the amount
of sunlight hours for longer and longer days.

Spring is gentle and sweet.
It offers lovely temperatures
and fragrant flowers.

Spring's job is to maintain healthy soil and easy weather conditions for new gardens, new baby animals, and new projects around our homes.

Spring's birthday is called Spring Equinox or Vernal Equinox.

There is an equal amount of daylight and nighttime darkness.

Spring Equinox SUPERCHARGES the LIGHT, giving us warmer and much longer days as it guides us to Summer.

Being outside all day fills me with excitement and happiness!

I love Spring!

I get to play soccer!

I make tiny spaces for fairies or gnomes around my yard.

Flowers and vegetables burst through the ground and start growing very quickly.

A Maypole Dance is a fun way to celebrate Spring!

The Maypole represents creating or recreating — starting anew, starting over, or starting again.

It symbolizes birth and rebirth.

After three months, Summer takes over from Spring.

Even if the weather still feels like Spring,
Summer will wave its magic wand to CHANGE THE LIGHT'S DIRECTION to go from long days to shorter days.

Summer is **HOT!**

Shade and water are our friends.

Summer's job is to heat the earth and create thunderstorms. This provides more energy for all living things.

It also allows roots and worms to move in the dirt, which keeps the soil fluffy.

This way, water can get through and crops can keep growing.

Summer's birthday is called Summer Solstice.

It's the longest day and shortest night of the year.

Summer Solstice activates the DECLINE of LIGHT!
Each day loses a bit more Sun time as we move toward Fall.

As the days get hotter and lose sunlight hours,
I begin to slow down.

The heat makes me feel lazy, but cooling off is fun!

My friends and I play
in the pool!

We eat ice cream.

Vegetables are growing and
some are ready to pick!

To beat the heat,
my parents take my
friends and me
camping by a river.

After three months, Fall takes over from Summer.

Even if the weather still feels like Summer,
Fall will wave its magic wand to turn the
SUNLIGHT OFF MUCH EARLIER in the evenings.

Fall is the last Season to visit.
It offers its magic to everything it touches!

Fall's job is to call on long cold nights
to bring an end to the year's growing cycle.

Leaves on trees change color and drop, and the
last of the garden vegetables are gathered.

Fall's birthday is called Fall Equinox.
It's also called Autumnal Equinox.

Like Spring Equinox, there is an equal amount
of daylight and nighttime darkness.

Fall Equinox SUPERCHARGES the DARK, increasing night
hours and shortening days as it takes us to Winter.

With the dark coming so early in the evening,
I get sleepy earlier.

I love Fall (and sleeping)!

We make lots of soup.
YUM!

I look forward to carving pumpkins every year.

It's time to harvest the rest of the fruit and vegetables that have been growing in the garden throughout the year.

Our community has a big gathering
called a Harvest Festival.

We cook a big meal and give thanks for the food on our table and our many blessings throughout the year.

After three months, Winter takes over from Fall.

Even if the weather still feels like Fall,
Winter will wave its magic wand to kickstart the
RETURN of LONGER DAYS of SUNLIGHT.

And the year starts all over again!

Fun facts for the family...

The Seasons serve as an important yearly clock for people, plants, and animals.

Throughout history, humans have moved with the Seasons to live in warmer or cooler conditions.

Creatures of all types such as birds, bugs, butterflies, and whales travel by land, sea, or air to reach their destination — to birth babies and to be closer to food sources.

The Seasons keep us connected to Nature and help all living things work together as a team to thrive.

The Seasons even guide you through at least FOUR different kinds of energy in one year!

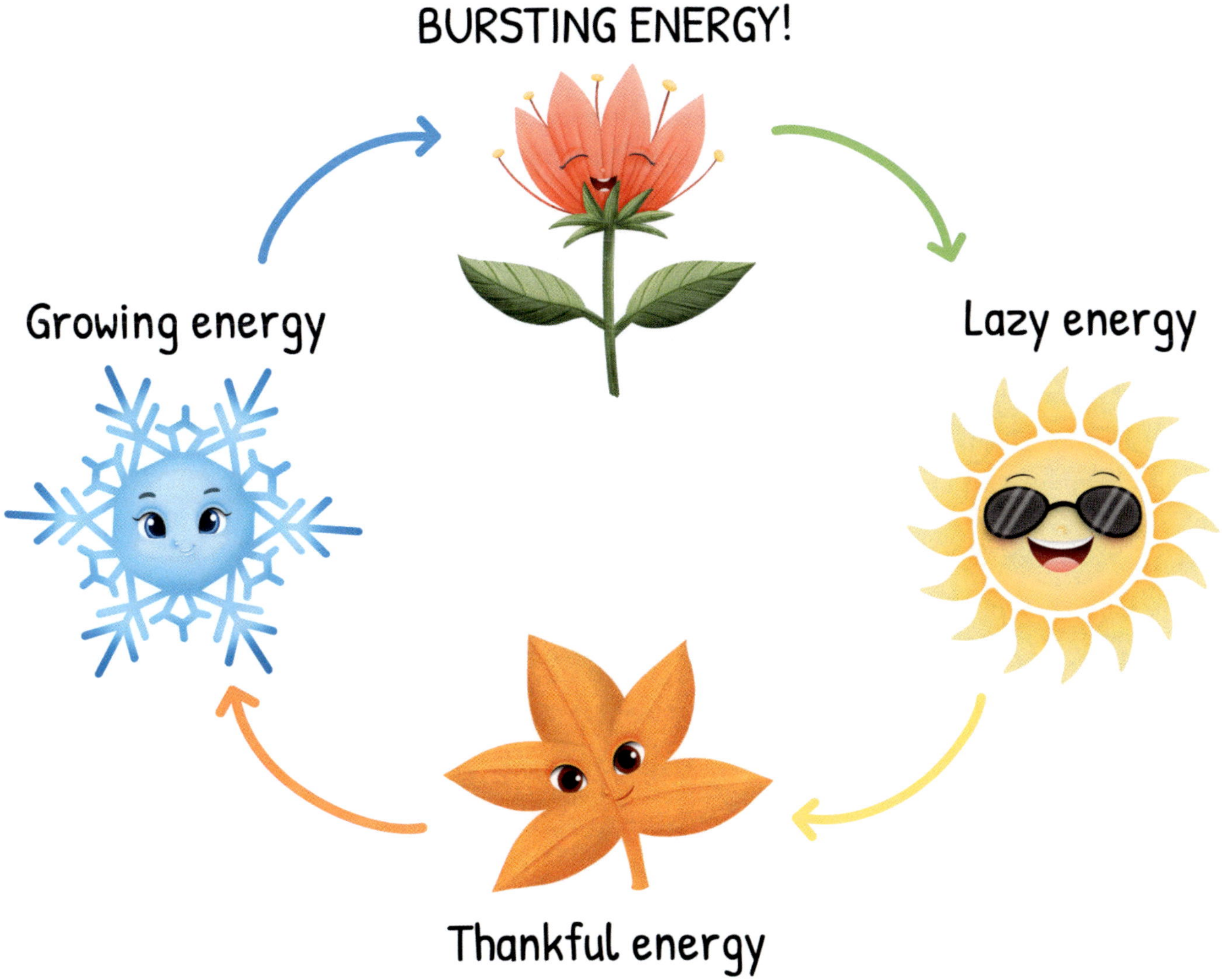

And then it does the same thing the next year, and the next year, and the next year!

This is called a "body rhythm."

There are many body rhythms inside of you.

The one that follows the seasonal light is called a "yearly body clock" or a "circannual rhythm."

sur-KAN-u-ahl means "around a year."

RITH-um means "movement in time."

Do all places have Four Seasons?

- The closer you live to the Equator, the more your Seasons will be the same.
- The farther north or south you live from the Equator, the more likely you'll experience the Four Seasons.
- The Northern Hemisphere is north of the Equator.
- The Southern Hemisphere is south of the Equator.
- The Northern and Southern Hemispheres experience OPPOSITE Seasons. When it is Winter in the Northern Hemisphere, it is Summer in the Southern Hemisphere!

Sun and Moon and Seasons

> During the Summer, the Sun travels high in the sky and the Moon travels low in the sky.

> During the Winter, the Sun travels low in the sky and the Moon travels high in the sky.

Questions

1. What are the Four Seasons like where you live?
2. Which is your favorite Season?
3. What is your energy like in Winter?
4. What is your energy like in Spring?
5. What is your energy like during Summer?
6. What is your energy like in Fall?

Project Ideas

1. Make cookies with shapes from the four seasons (such as a snowflake, leaf, flower, and sun).

2. Draw pictures of what the Seasons look like where you live.

3. Make a seasonal "family fun list"! For each Season of the year, write down your favorite foods, outdoor activities, arts & crafts, events, games, etc.

4. Give someone you love a seasonal nature gift!
   - Winter: Tie small sticks together with a red ribbon.
   - Fall: Find a large or colorful leaf.
   - Spring: Collect a small bouquet of wild flowers.
   - Summer: Pick up a pretty pebble or shell.

Have fun with the Seasons!

A Note from the Author

The Seasons and Me is the third book in a series that introduces children and their people to the connection between Nature's rhythms and their own body rhythms. Learning to synchronize their daily, monthly, and yearly habits with the Sun, the Moon, and the Seasons may help families achieve ideal lifelong health and happiness!

This book shows optimal seasonal conditions for a healthy planet, but we can still tune into the Seasons when conditions are not usual. Our bodies are cued by our local daylight hours, which the Solstices and Equinoxes activate every three months despite climate change. Please note that human-created seasonal holidays and events, school schedules, and work hours are not always in alignment with what would be best for our biological bodies & mental health. The yearly solar clock guides & benefits our vitality, motivation, and driving spirit each day! Aligning with the Seasons creates an annual pattern and routine that supports a positive connection to all around us.

This book tells *that* story in a way that's easier to understand!

I encourage your family to notice how each Season's *light* affects your actions and to incorporate the conversation into your daily and yearly language. Large words used in the book provide good teaching opportunities and discussions. Words not normally capitalized are done so because they are considered the series' main characters.

The Sun and Me, The Moon and Me, and The Seasons and Me teach body rhythm information to kids of all genders. The *Connect with Nature* series continues with books to help those with girl bodies to know and love the superpowers of their monthly hormonal rhythms beginning at adolescence, based on the *4 Seasons in 4 Weeks* cycle-tracking method.

Suzanne Mathis McQueen

Suzanne is the author of the book *4 Seasons in 4 Weeks*.
She is a Nature nerd who hugs trees and
chats with ladybugs at her home in Southern Oregon.
Fall is her favorite season.

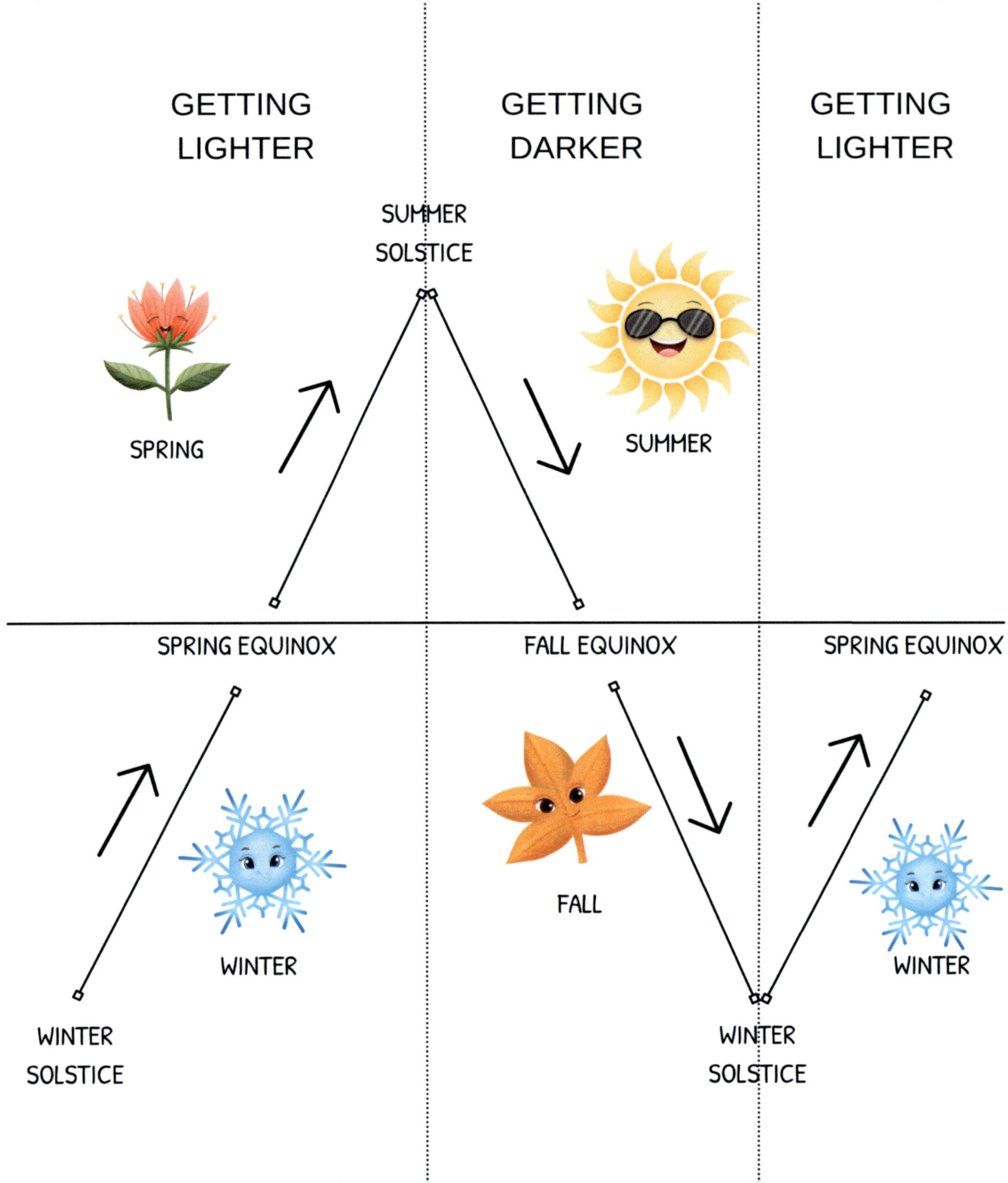
GETTING LIGHTER
GETTING DARKER
GETTING LIGHTER
SUMMER SOLSTICE
SPRING
SUMMER
SPRING EQUINOX
FALL EQUINOX
SPRING EQUINOX
FALL
WINTER
WINTER
WINTER SOLSTICE
WINTER SOLSTICE

Have you read all of these books by Suzanne Mathis McQueen?

The Connect with Nature Series

For children & their people about body clock awareness for health & happiness!

and...Coming Soon!

A coming-of-age picture book for 7-11- year-old girls!

For women, couples, and mature teens:

Based on science, the *4 Seasons in 4 Weeks* cycle-tracking method identifies an approximate 28-day female body clock and breaks it down into four non-technical guided weeks for optimal decision-making in health, business, and love.

Includes women's history, sacred geometry, tools and self-care guidance for each cycle week, information for men, and guidance for not-cycling women as well. Explicit content with 300 pages of full color, gorgeous artwork, and color-coded sections.

Reviews are appreciated...

If you enjoy reading *The Seasons and Me* and believe other kids (and their people) would also like this book, please leave a review on your favorite review site! Reviews help readers find good books and independent publishers (like us!) to stay in business.

Thank you!

Stay in touch...
Scan this code or type in the URL below to find coloring pages, contact info, merch, and other bits of fun:

4s4w.com/connect-with-nature-books/

Made in the USA
Monee, IL
06 December 2025

37642450R00036